MW01130876
This book belongs to:

Our Earth is the only planet we know about that has water to drink. This makes it special.

1
2
3
4
5
6
7
8
9
10
11
12
13
14
15
16

An atom is a basic unit of matter. It has a nucleus in the center with a cloud of electrons around it.

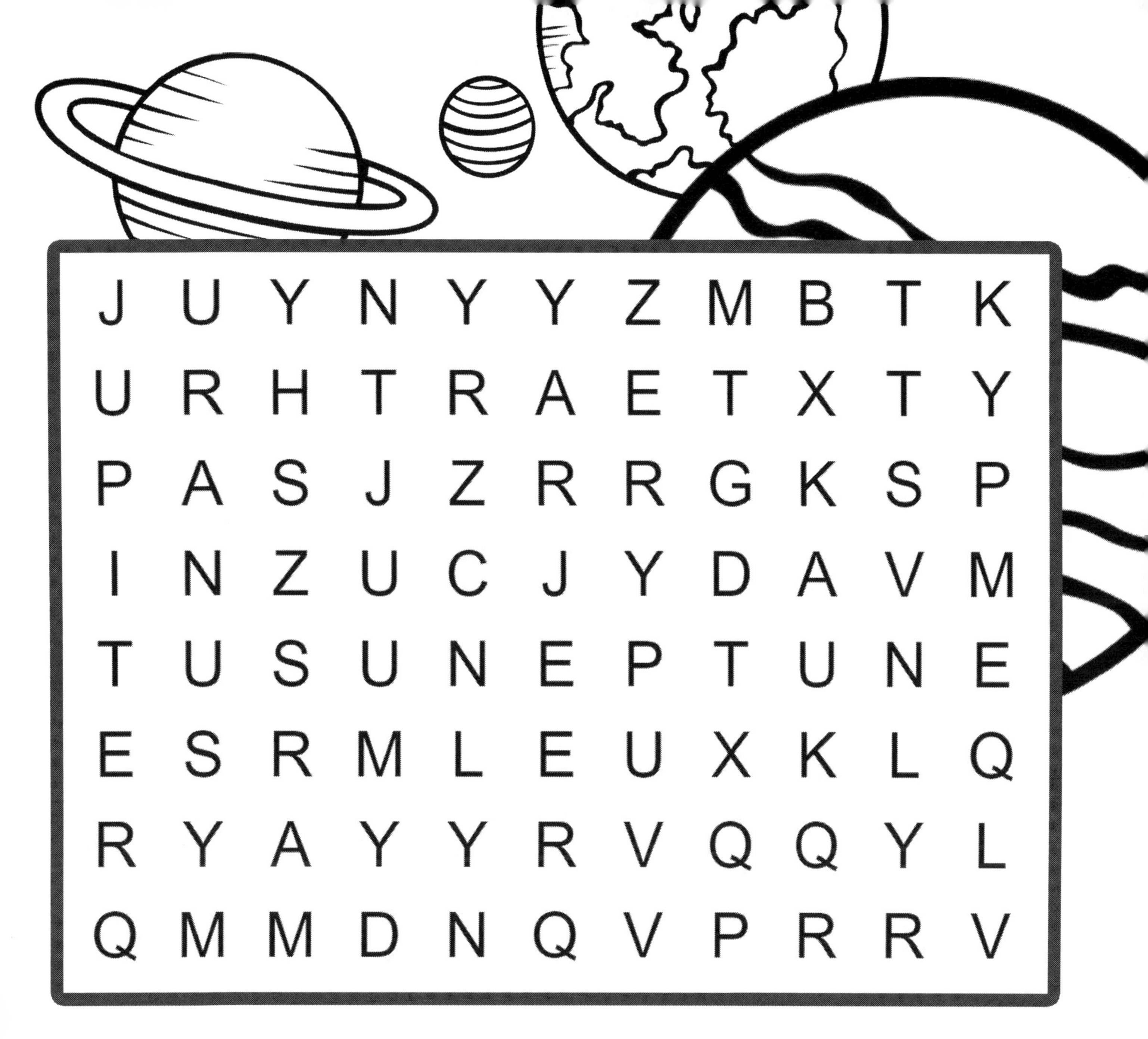

MERCURY - the smallest planet in our solar system

VENUS - the hottest planet in our solar system

EARTH - the only planet with liquid water in our solar system

MARS - scientists believe Mars used to be wetter and warmer than now

JUPITER - more than twice as big than the other planets combined

SATURN - has large rings around it

URANUS - the 7th planet from the sun

NEPTUNE - the farthest planet from the sun and is dark and cold

magnet

A magnet is a piece of rock or metal that can pull some types of metal toward itself.

ESCAPE
TO THE
STARS

Insect Anatomy

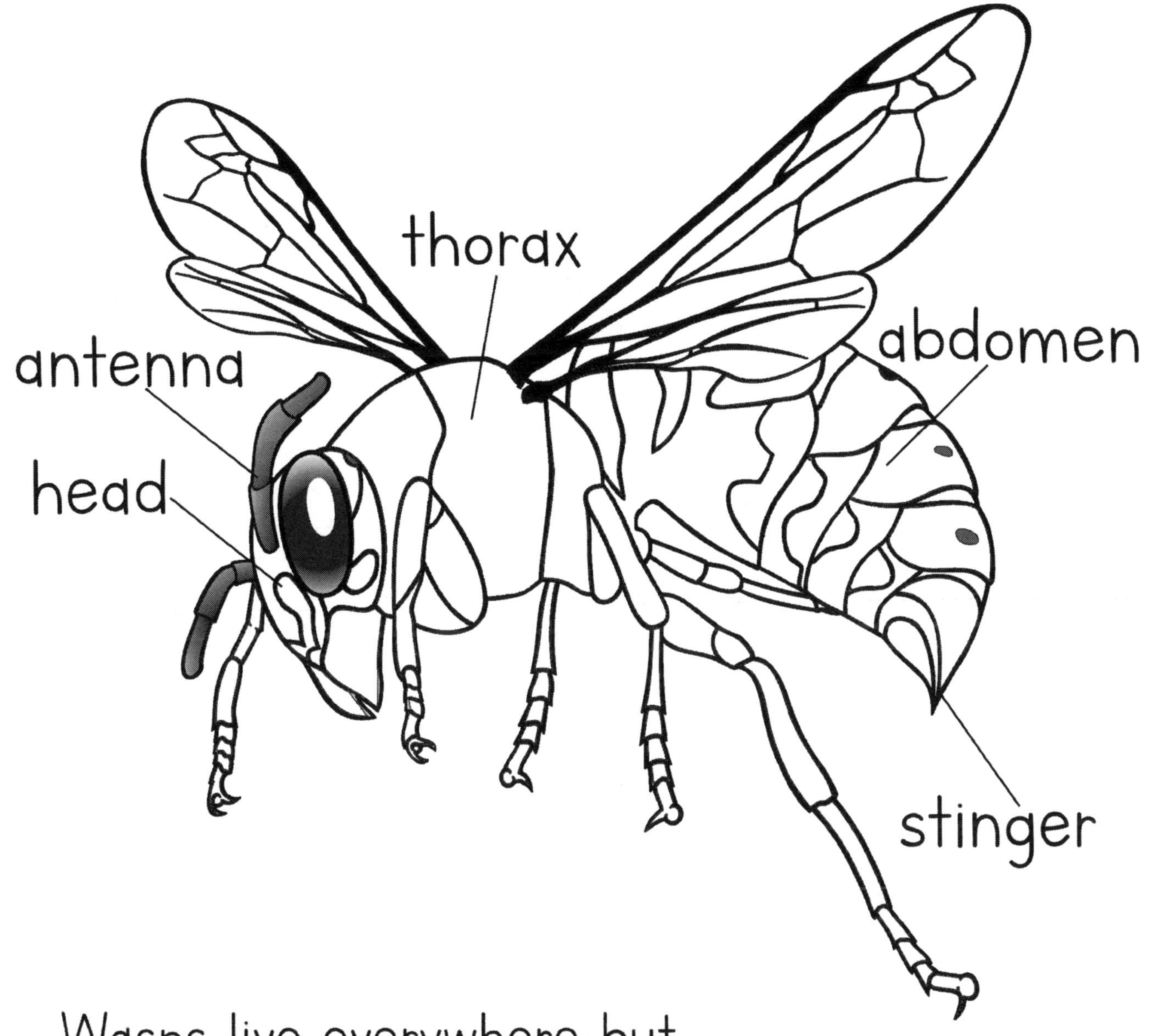

Wasps live everywhere but Antarctica. They chew up bark and spit it out again to form a rough paper. This is what they make their nests out of.

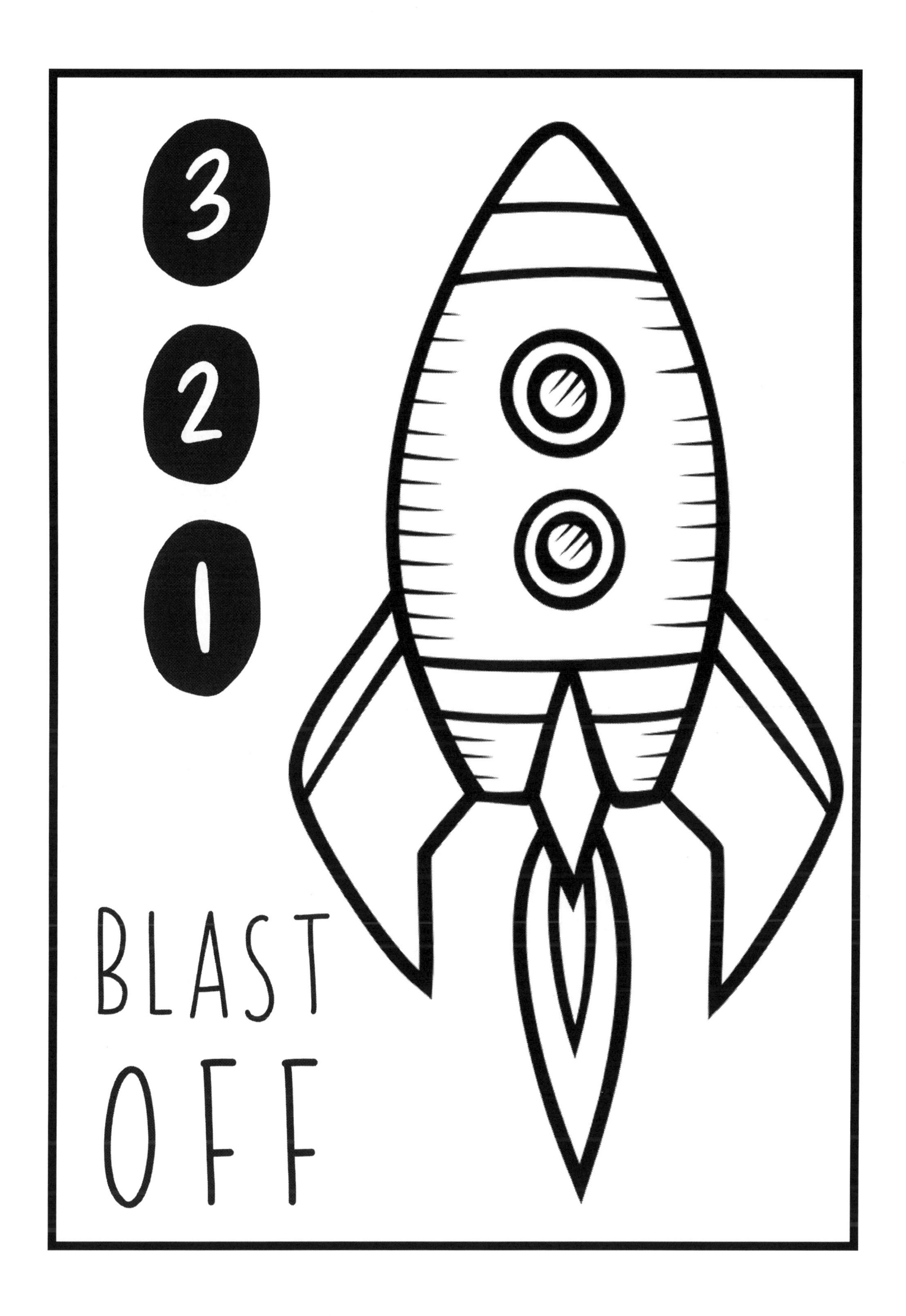
3
2
1
BLAST
OFF

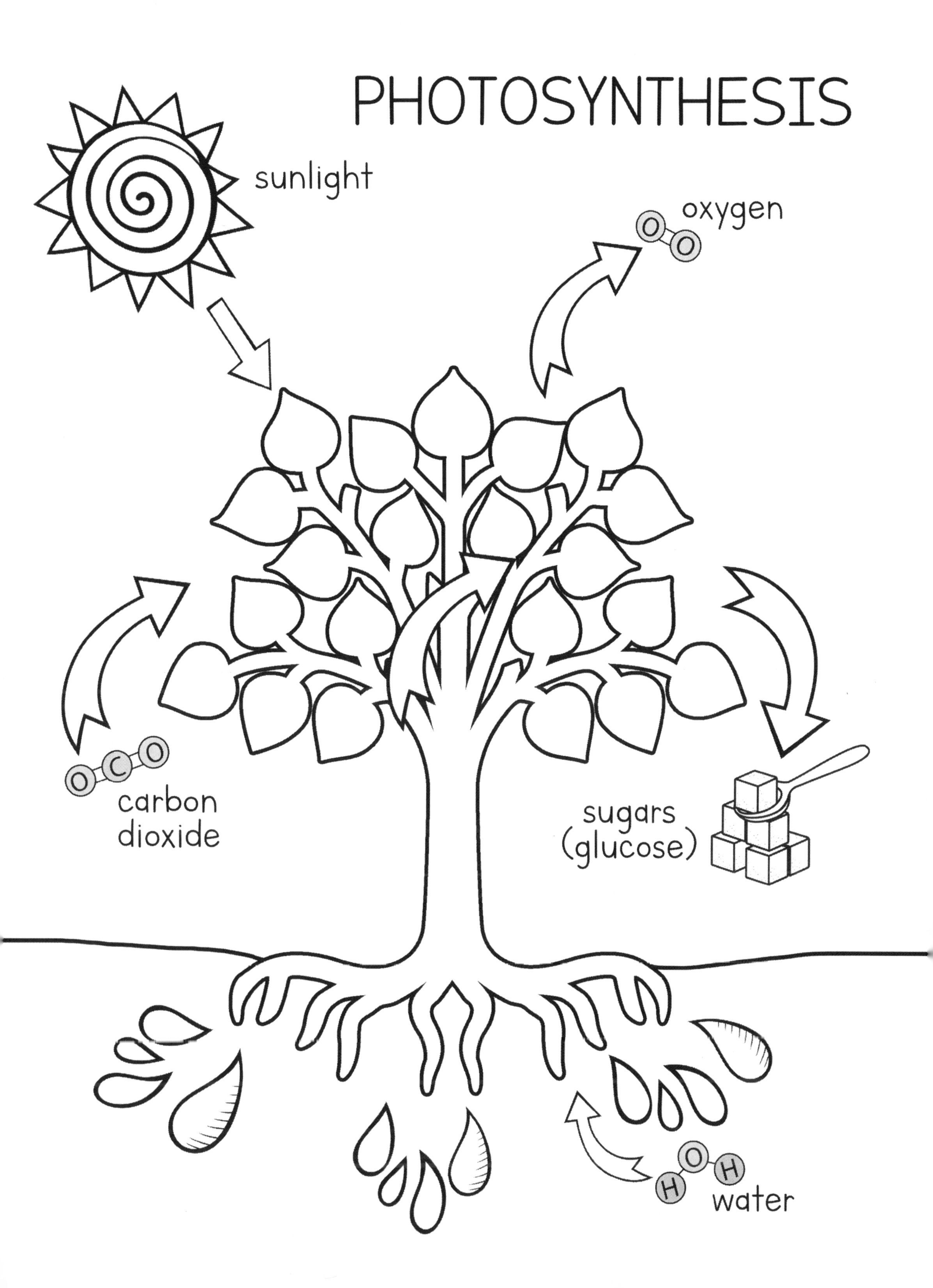
PHOTOSYNTHESIS
sunlight
oxygen
O O
O C O
carbon
dioxide
sugars
(glucose)
H O H
water

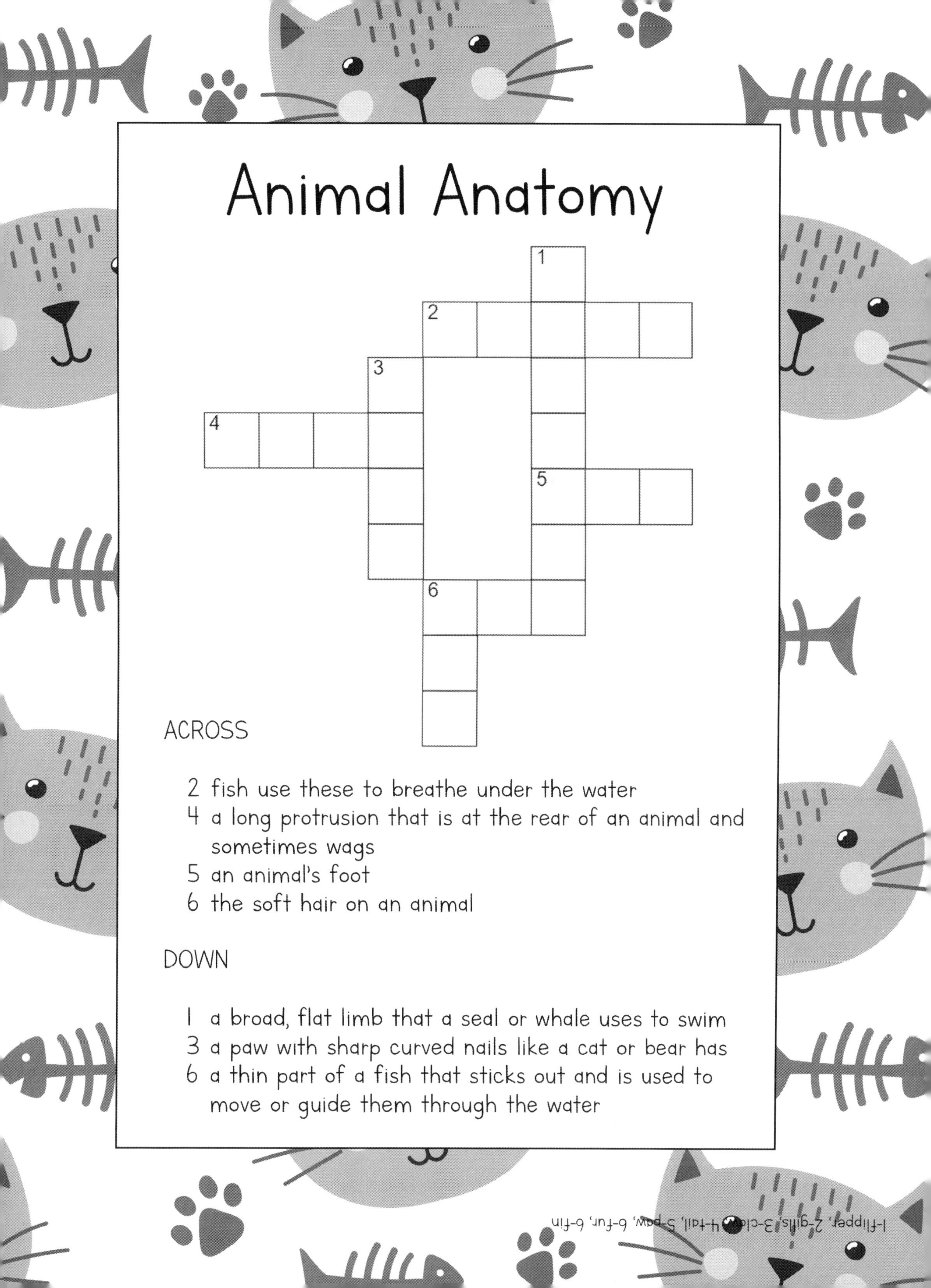

Animal Anatomy

ACROSS

2 fish use these to breathe under the water
4 a long protrusion that is at the rear of an animal and sometimes wags
5 an animal's foot
6 the soft hair on an animal

DOWN

1 a broad, flat limb that a seal or whale uses to swim
3 a paw with sharp curved nails like a cat or bear has
6 a thin part of a fish that sticks out and is used to move or guide them through the water

1-flipper, 2-gills, 3-claw, 4-tail, 5-paw, 6-fur, 6-fin

organism

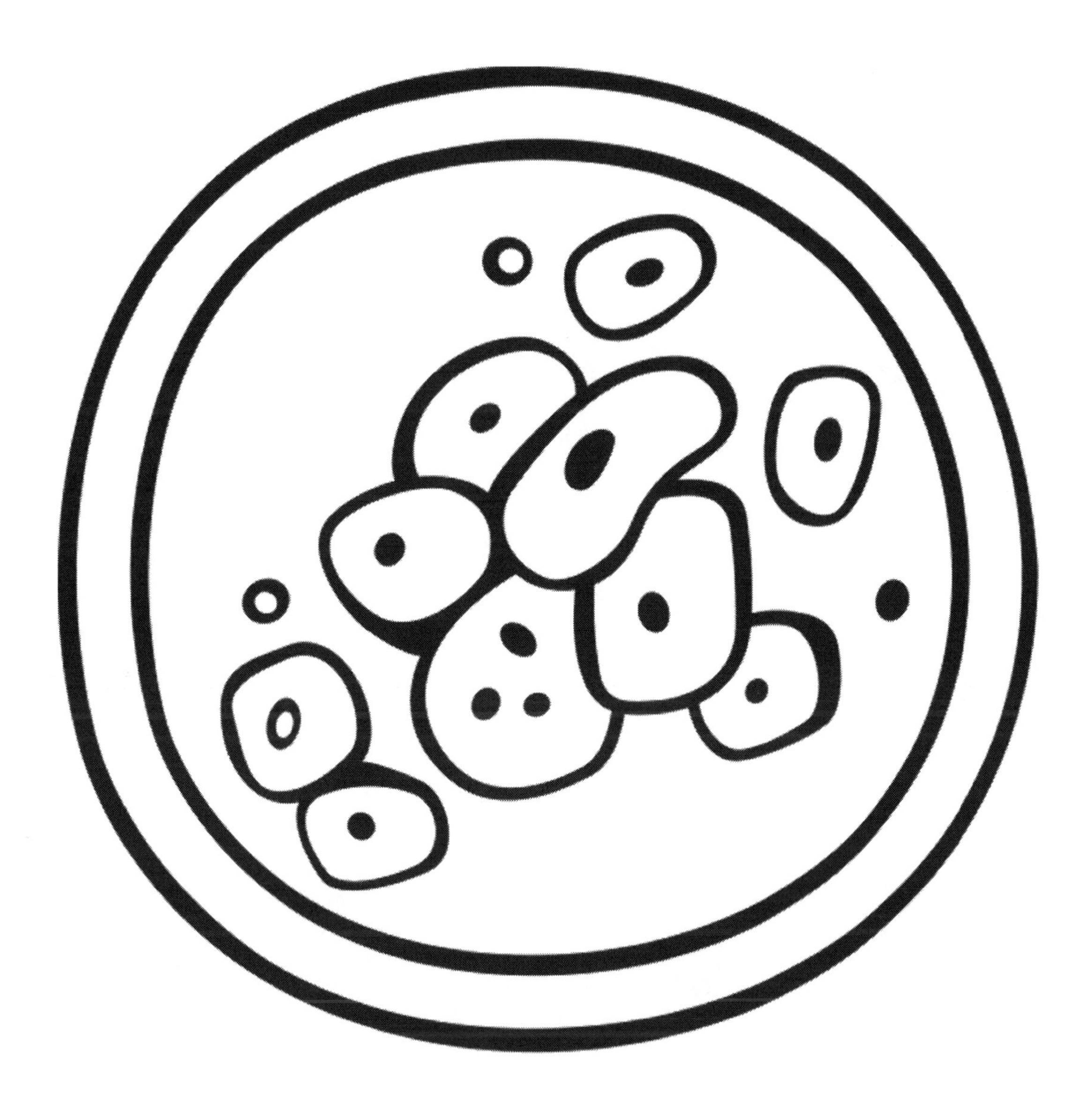

An organism is a living thing made up of one or more cells and able to carry on the acitivies of life such as using energy, growing, and reproducing.

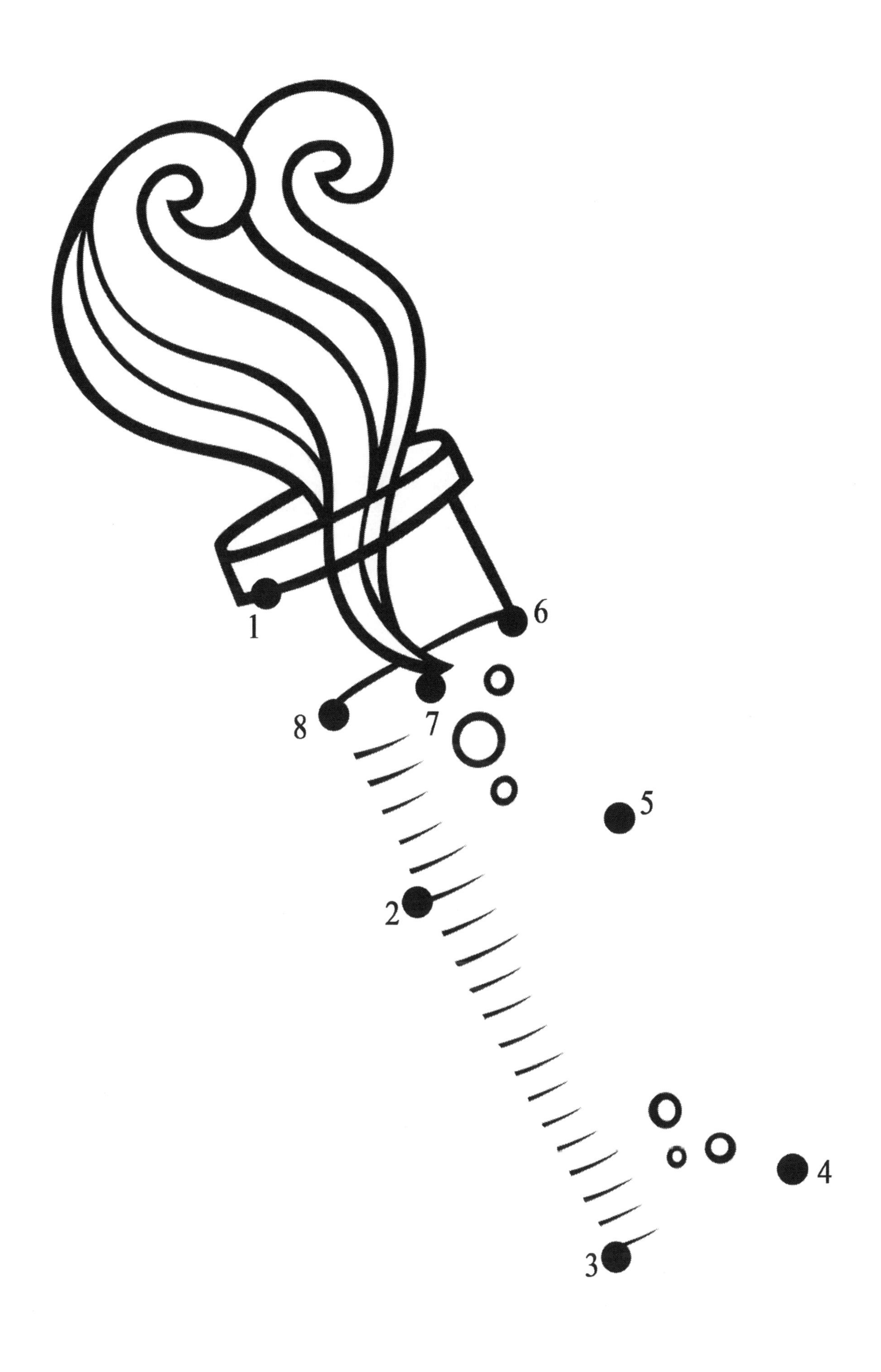
1
6
8
7
5
2
4
3

Leaf Anatomy

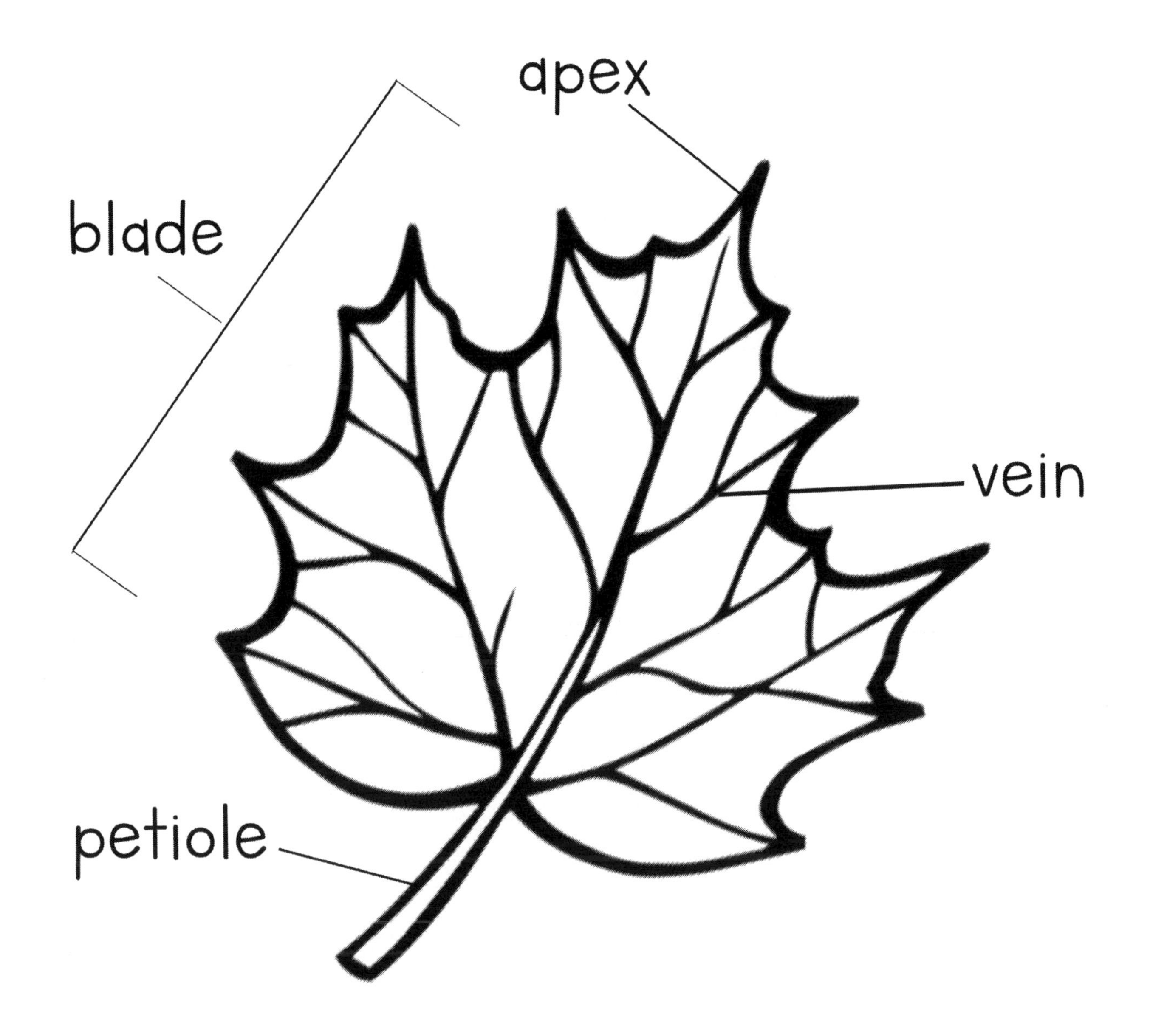

A leaf's main functions are photosynthesis and gas exchange. They take in carbon dioxide and they release oxygen.

A microscope is used to look at tiny objects like cells from an animal or plant. It can make things look several hundred times bigger than they really are so you can see all the little parts clearly.

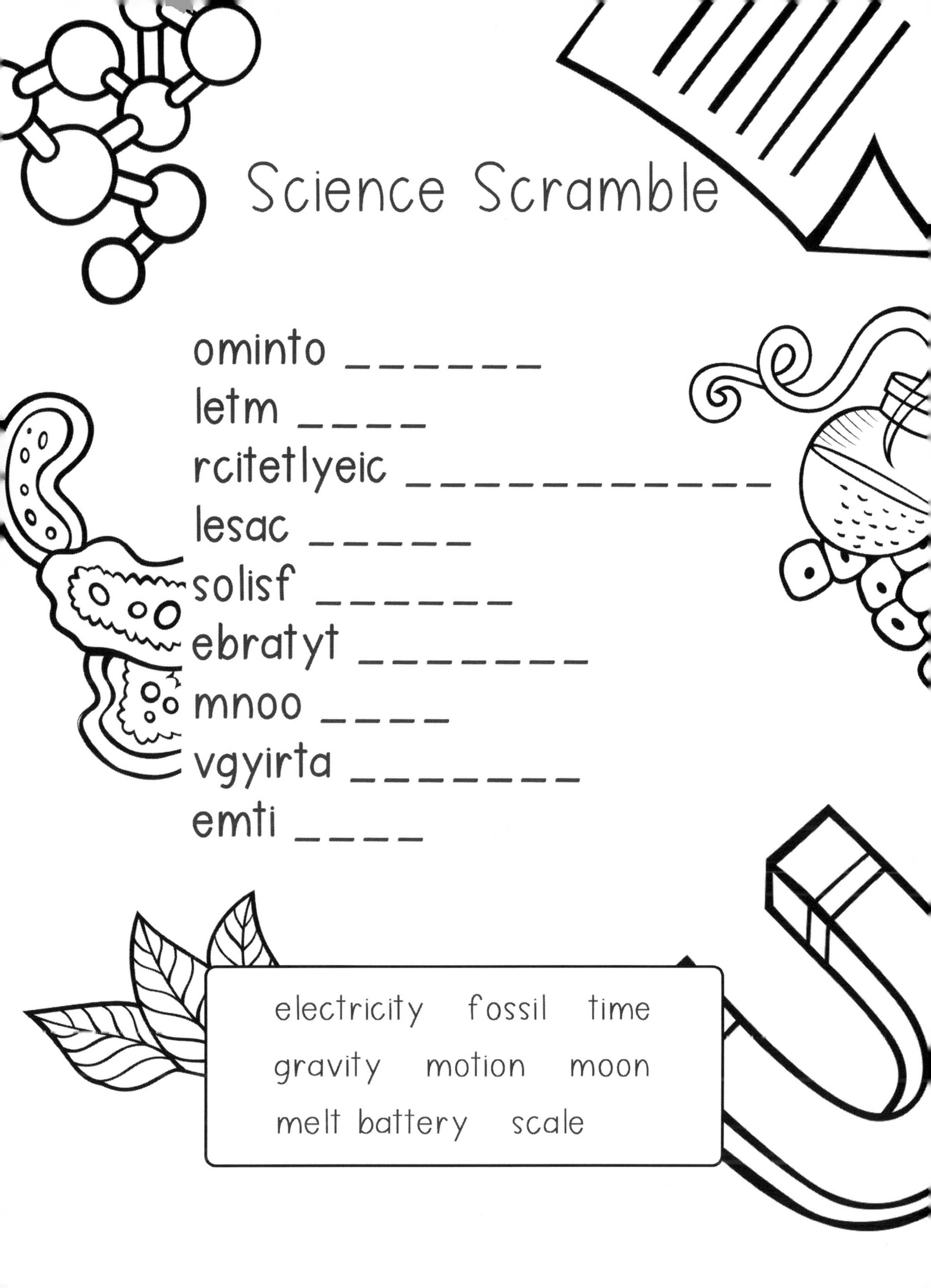

Science Scramble

ominto _ _ _ _ _ _

letm _ _ _ _

rcitetlyeic _ _ _ _ _ _ _ _ _ _ _

lesac _ _ _ _ _

solisf _ _ _ _ _ _

ebratyt _ _ _ _ _ _ _

mnoo _ _ _ _

vgyirta _ _ _ _ _ _ _

emti _ _ _ _

electricity fossil time

gravity motion moon

melt battery scale

1 2 3 4 5

ACROSS

4 when it does not rain for a long time causing life that needs water to suffer

5 a lizardlike reptile that lived millions of years ago but is now extinct

DOWN

3 The preserved remains or traces of an animal or plant that lived a very long time ago

1 a type of plant or animal that no longer lives anywhere

2 a mountain that erupts hot ash and lava

1-extinct, 2-volcano, 3-fossil, 4-drought, 5-dinosaur

molecule

A molecule is the smallest unit of a substance that has all the properties of that substance. For example, a water molecule is the smallest unit that is still water. Molecules are made of atoms.

EXPLORE!

amoeba

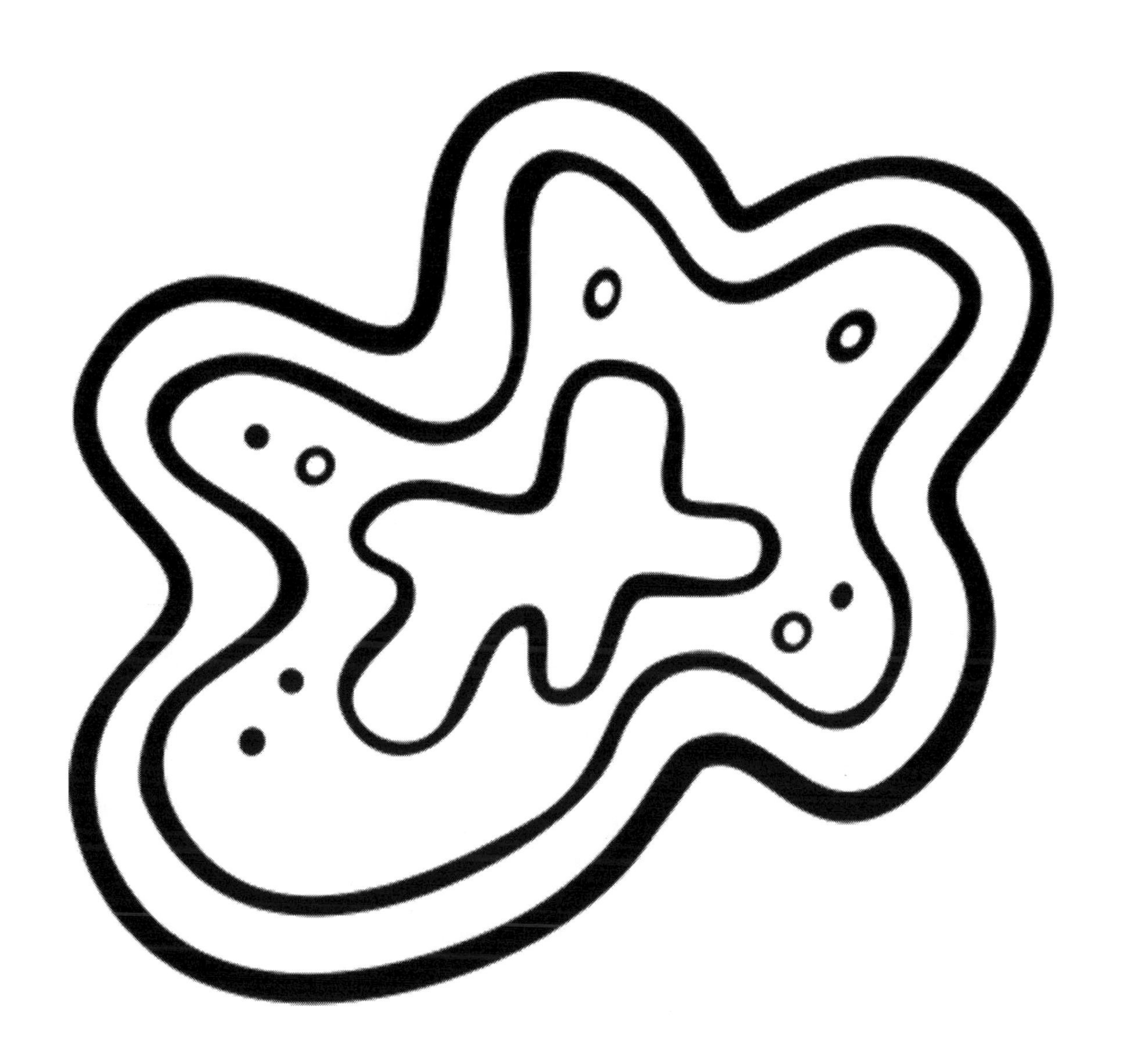

An amoeba is a tiny creature made up of one cell and is among the simplest of living organisms.

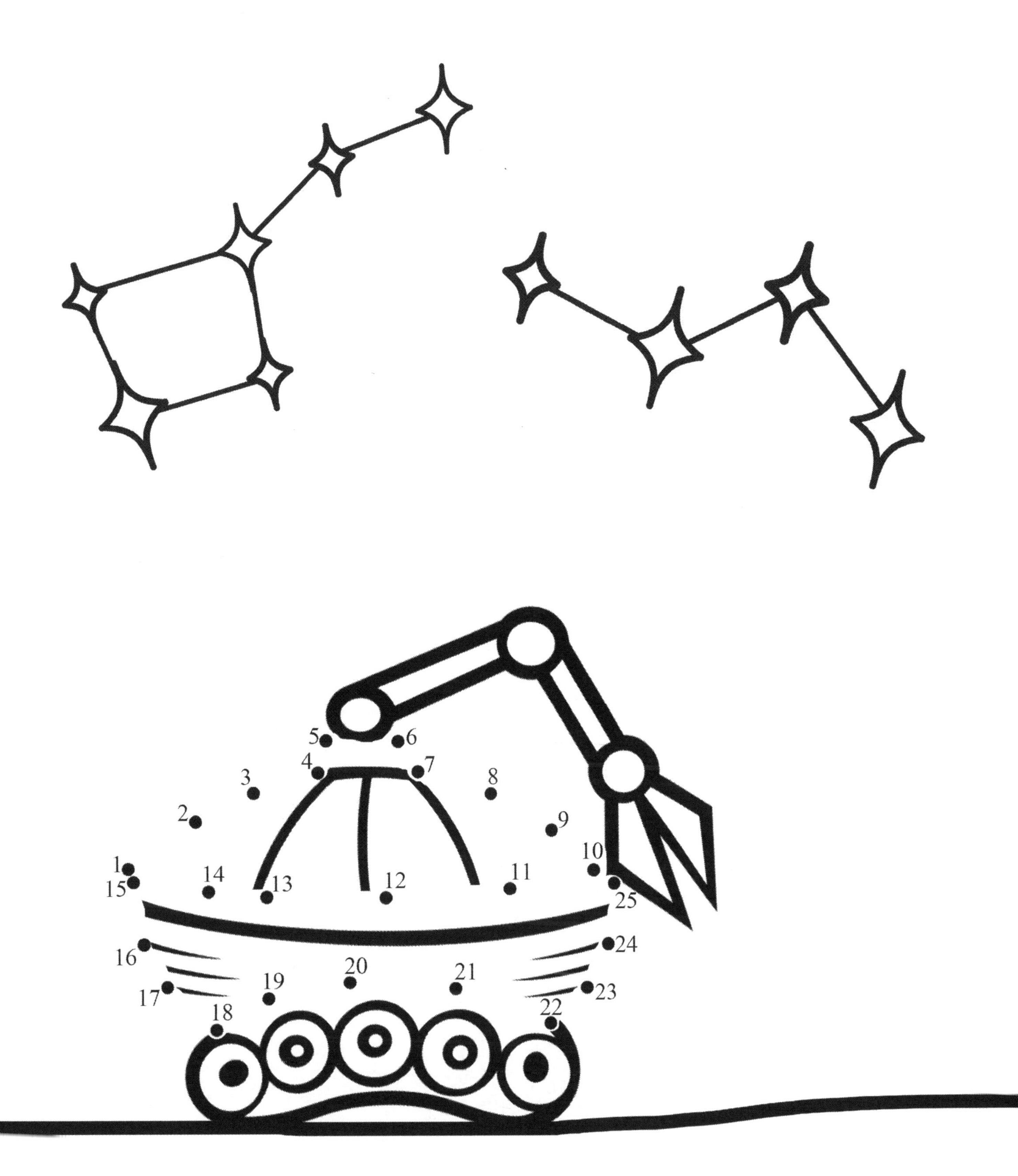

A rover is a machine made to land on another planet or moon to explore, take pictures, and collect information.

Earthworm Anatomy

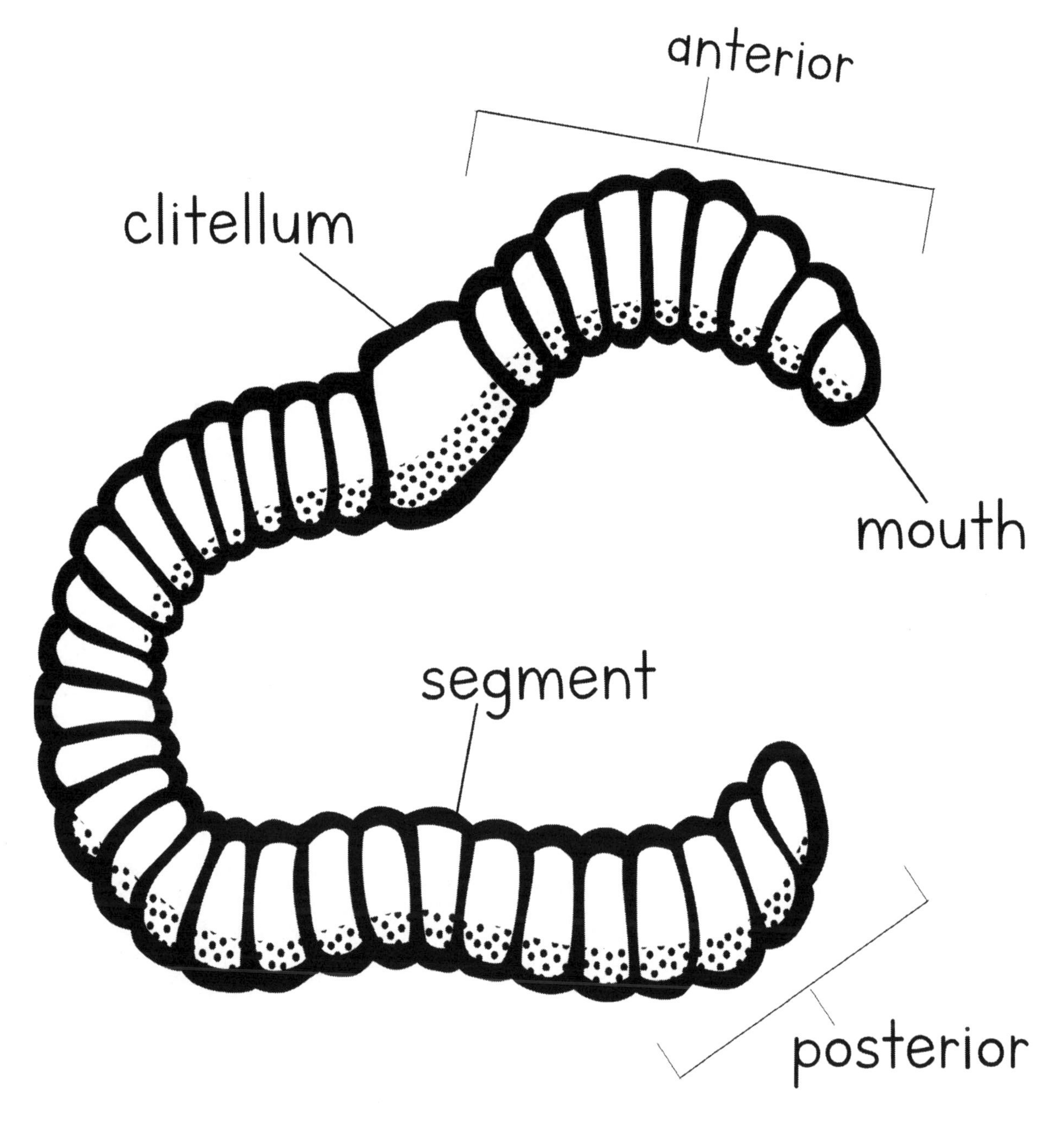

Earthworms are made up of segments. They move minerals and nutrients in the soil around and help make the soil healthy for plants. They can eat up to a third of their body weight in one day.

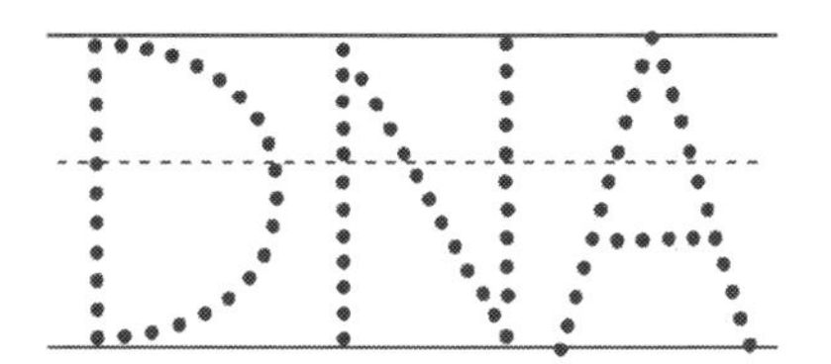

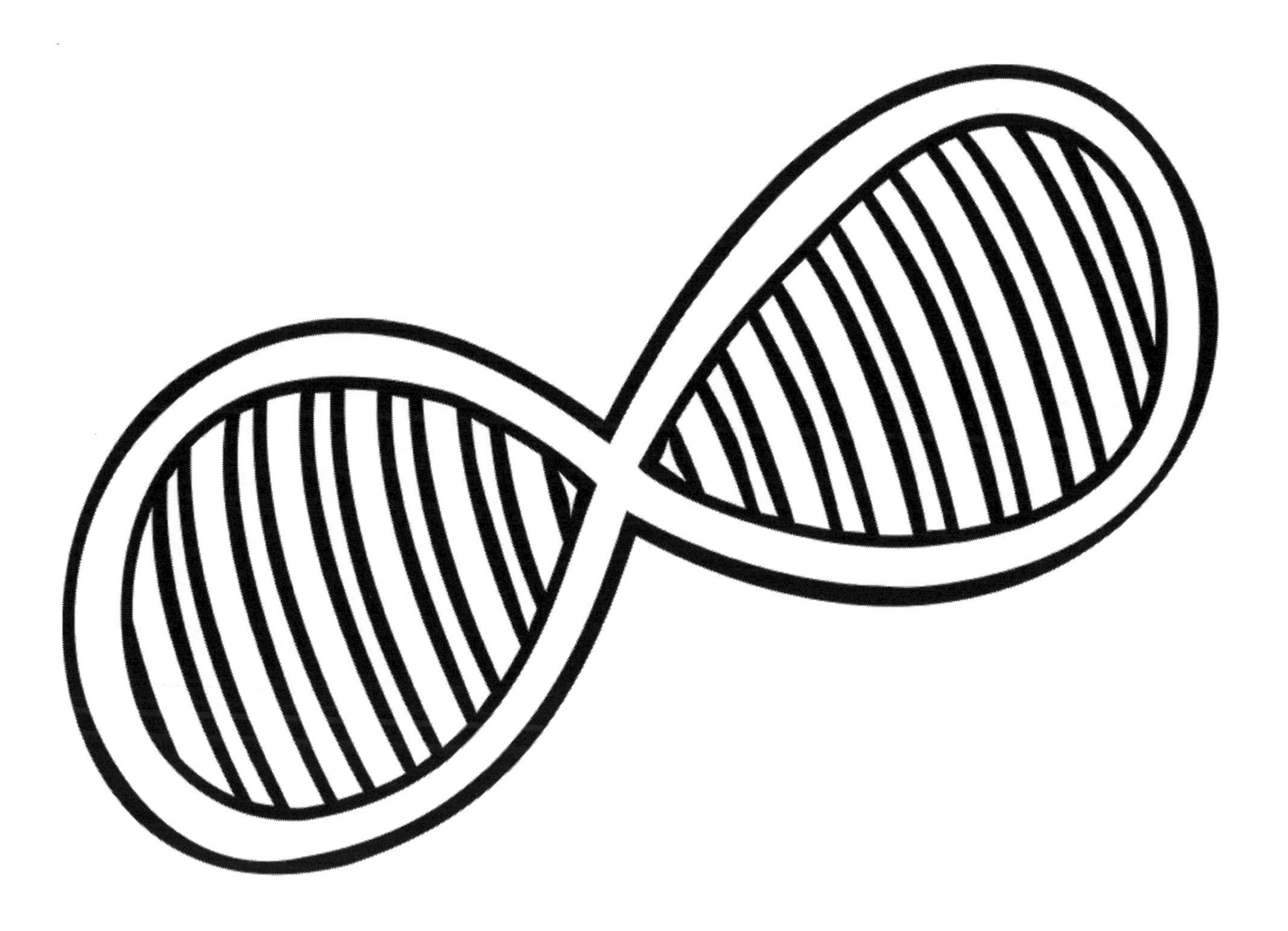

DNA is in every cell of living things. DNA carries all of the information about how a living thing will look and work. In people, DNA determines how tall you will be and what color eyes you have.

H Q B Q T I B R O W
E G J B E H P K Y L
A J A W N Z G D B Y
T W X S A W D I A N
R N Z N L L E R N Y
J K O M P A A Z X R
T O T S R T Y Q Q R
M P K T S W X Z Y W
T Y H Y T T R T N N
STAR
PLANET
EARTH
ORBIT
NIGHT
SKY
DAY
GAS
MOON
HEAT

Anatomy Scramble

tppateei _ _ _ _ _ _ _ _ _

tooht _ _ _ _ _

iskn _ _ _ _

waj _ _ _

gusln _ _ _ _ _

ese _ _ _

seens _ _ _ _ _

rhae _ _ _ _

hbetera _ _ _ _ _ _ _

sense	skin	hear
jaw	appetite	lungs
breathe	see	tooth

satellite

An artificial satellite is an object made by humans that orbits (or goes around) the Earth. Satellites are sent to gather information, take pictures, or beam signals for television or phone calls.

Made in the USA
Columbia, SC
23 March 2022